Body Needs

FATS
for a healthy body

Heinemann Library
Chicago, Illinois

Jillian Powell

Created by the publishing team at Heinemann Library
Designed by Ron Kamen and Celia Floyd
Illustrations by Geoff Ward
Originated by Ambassador Litho
Printed in China by Wing King Tong

07 06 05 04 03
10 9 8 7 6 5 4 3 2 1

Library of Congress Cataloging-in-Publication Data
Powell, Jillian.
    Fats for a healthy body / Jillian Powell.
      p. cm. -- (Body needs)
    Summary: Discusses what fats are, how they are absorbed and stored in the body, how the body uses fats, and health problems caused by fats.
    Includes bibliographical references and index.
    ISBN 1-40340-757-6 (lib. bdg.) ISBN 1-40343-311-9 (pbk.)
    1. Fatty acids in human nutrition--Juvenile literature.
    2. Lipids in human nutrition--Juvenile literature.
    [1. Fat. 2. Nutrition.]
    I. Title. II. Series.
    QP751.P695 2003
    612.3'97--dc21

2002012643

Acknowledgments
The author and publishers are grateful to the following for permission to reproduce copyright material: p. 4 Anthony Blake; p. 5 Steve Behr; pp. 7, 32 Gareth Boden; pp. 9, 21, 22, 27, 38 Liz Eddison; pp. 13 (Chris Priest and Mark Clarke), 14, 26 (Will and Deni McIntyre), 35 SPL; pp. 16, 36, 37 (Bohemian Nomad Picture Makers), 39 (Stock market/Ariel Skelley), 41 (Macduff Everton), 43 Corbis; p. 19 (Reinhard Krause) Reuters; p. 24 Trevor Clifford; pp. 25, 28 Photodisc; pp. 29, 33 Getty Stone; p. 31 Neil Phillips.

Cover photograph of butter, reproduced with permission of Gareth Boden

Every effort has been made to contact copyright holders of any material reproduced in this book. Any omissions will be rectified in subsequent printings if notice is given to the publisher.

Some words are shown in bold, **like this.** You can find out what they mean by looking in the glossary.

Contents

Why Do We Need to Eat?

Most people eat two or three main meals a day. We eat because we get hungry and because we enjoy the taste of food. At the same time we are giving the body the chemicals it needs to stay alive and healthy.

Cells

Your body is made up of millions of tiny **cells.** For example, your bones consist of bone cells and your skin is made of skin cells. Most cells are so small you need a microscope to see them, but they are all working hard to carry out various jobs. For this reason, your cells need a steady supply of **energy.** They also need many different substances, which come mainly from the food you eat. These substances are called **nutrients.**

Nutrients

Carbohydrates, fats, **proteins, vitamins,** and **minerals** are all different kinds of nutrients. Most foods contain a lot of one kind of nutrient, plus small amounts of other nutrients. Together, nutrients provide energy and materials that the body needs to work properly and to grow. This book is about fats. You will learn what they are and how the body uses them. We will also explore other nutrients and how they work with fats to make you healthy.

Whether you eat your food at home, at school, or in a restaurant, you should try to eat a balanced and healthful meal.

Energy food

Your body's main need is for food that gives you energy. Everything you do uses energy. You use energy when you run or swim. You also use energy when you think, eat, and sleep. Carbohydrates and fats provide energy. The body burns carbohydrates, just like a car engine burns gasoline. Your body needs a big supply of energy every day. Foods such as bread, pasta, potatoes, and sugar are carbohydrates that are your body's main source of energy.

Protein

You need protein to make new cells and repair any damaged ones. Protein is the main substance found in muscles, skin, and the **organs** inside your body. Your body is constantly making new skin cells, muscle cells, and other types of cells. Cells consist mainly of water and protein, so to build new cells your body uses proteins that come from foods such as meat, fish, eggs, beans, and cheese. It is especially important that children eat plenty of protein, because they are still growing, and their bodies need it to make millions of extra cells.

When you exercise or play sports, your body uses the energy you get from food.

What Are Fats?

The food we eat helps our bodies grow, stay healthy, and have **energy**. For a balanced diet, we need to eat a wide range of foods that contain **carbohydrates, proteins,** and fats.

Energy storage

Like carbohydrates, fats and oils provide energy, which is stored in the foods we eat. The white streaks you see in a steak or a piece of bacon are fat. All animals use fat to keep warm and to store energy. The oil inside of nuts and grains is a stored form of energy for plants. Oils are fats that are liquid at room temperature.

We eat fats in lots of different foods. We eat animal fats in meat and dairy foods such as butter, milk, and cheese. We eat vegetable oils when we use corn oil or olive oil to fry foods or make salad dressing. Fats are also used in home-baked foods such as cookies or cupcakes and in packaged foods such as potato chips and crackers.

Corn Oil
The oil in a kernel of corn comes from inside the **germ**. If you cut a corn kernel in half, you can see the **bran, starch,** and germ. The oil can be pressed out of the germ. It takes fifteen ears of corn to make half an ounce (15 milliliters) of corn oil.

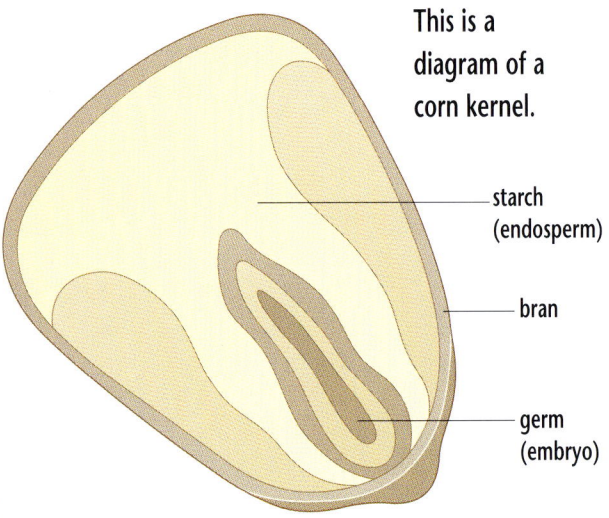

This is a diagram of a corn kernel.

starch (endosperm)

bran

germ (embryo)

Do you need fats?

Fats are sometimes seen as being bad for you because they are high in **calories.** If you eat too much fat, you may gain weight. Too many fried or other fatty foods will increase your risk of becoming overweight or **obese.** But not all fats are bad for you. You need some fats in your diet to keep your body healthy.

You need fats in the diet to give you energy and help you grow. Having fats in the diet also helps your body absorb **vitamins** A, D, E, and K. Fat stored in the body helps keep you warm and cushions and protects organs, such as your **liver.**

Fatty acids

Foods that contain fat also have **fatty acids** that your body needs to stay healthy. Your body cannot make these fatty acids. You can only get them from fats that you eat. You need them to keep your brain and nerve **cells** healthy. They make skin oils and help form chemicals called **hormones** that your body needs in order to work right. They also help your body fight off germs and diseases and repair damaged tissue.

Fats can also make food taste better and improve its texture. One reason cakes, doughnuts, and cookies are tasty is because they have a lot of fat in them. However, eating too many fatty foods, which provide lots of energy, may lead to weight gain.

Foods that contain fats give you lots of energy.

Fats and Flavor
Foods such as french fries, potato chips, onion rings, and fried chicken are tempting to eat because fats add flavor and texture to food. Frying foods in fat gives them a crispy texture.

What Is in Fats?

All the fats and oils in foods that we eat are made out of **fatty acids.** There are many different types of fatty acids, but they are all made from the same basic substances. A fatty acid is always a chain of three different **elements: carbon, oxygen,** and **hydrogen.** Different fatty acids have different amounts of these three elements.

Foods that contain fats and foods that contain **carbohydrates** both provide **energy** when we eat them. But fats are a richer source of energy than carbohydrates. Each 0.035 ounce (1 gram) of fat provides 9 **calories** of energy—more than twice as many as the same amount of carbohydrate.

Fatty acids

There are more than 40 types of fatty acids in foods and about 21 in the average diet. Three main types of fatty acids are **saturated, polyunsaturated,** and **monounsaturated.** Scientists tell the different types apart by counting how many hydrogen **atoms** they have.

Saturated fatty acid has the greatest possible number of hydrogen atoms attached to every carbon atom. Scientists say it is "saturated" with hydrogen, because it cannot hold any more. This type of fatty acid is mainly found in animal foods such as meat and cheese.

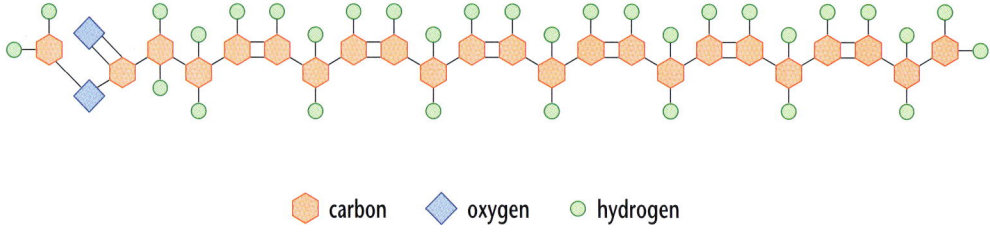

⬡ carbon   ◆ oxygen   ○ hydrogen

This is a fatty acid chain in a molecule of unsaturated fat. A whole molecule is made up of three chains.

Fat Fish Fact
Salmon and herring are fish that live in cold waters. They are high in polyunsaturated fats. Saturated fat is solid in cold temperatures. If the fish were high in saturated fats, they would freeze solid in icy water!

Find the Fats

A sample of food can be tested for fats by shaking it in a special type of alcohol called ethanol. Any fats **dissolve** in ethanol. This **solution** can then be poured into water. Fats are **insoluble** in water, so tiny droplets of fats will form in the water if they are present. A simpler test of fats is to press a sample of food between two sheets of paper. Fats and oils will leave a greasy mark on the paper.

Some fatty acids have one fewer pair of hydrogen atoms than saturated fats. They are called monounsaturated fatty acids. (*Mono* means "one.") These are found in foods such as olive oil and peanuts. Fatty acids that have even fewer pairs of hydrogen atoms are called polyunsaturated fatty acids. (*Poly* means "many.") They are found in vegetable oils like corn oil.

Saturated fats such as butter have a melting temperature of about 86 °F (30 °C), so they are solid at room temperature. Monounsaturated fats like olive oil are liquid at room temperature. They turn cloudy and begin to thicken when they are kept in cold temperatures. Polyunsaturated fats such as corn oil stay liquid even in the refrigerator.

Fats can be solid or liquid depending on the type of fatty acids they contain.

How Do We Get Fats from Food?

When you eat foods containing fats, your body needs to break the fats down so it can use them for **energy** and other body needs. Fats are broken down by your **digestive system.**

Eating and digesting

First you chew food, mashing it up and mixing it with **saliva** so it is softer and easier to swallow. Then the food passes down the esophagus into your stomach where stomach acids and **enzymes** start to break it down. Enzymes are a kind of **protein.** Their job is to speed up **chemical reactions** in **cells.**

Your stomach begins to digest proteins from your food, and then passes it on to your small intestine. The small intestine's job is to break food down into **soluble** particles that can be absorbed into your blood.

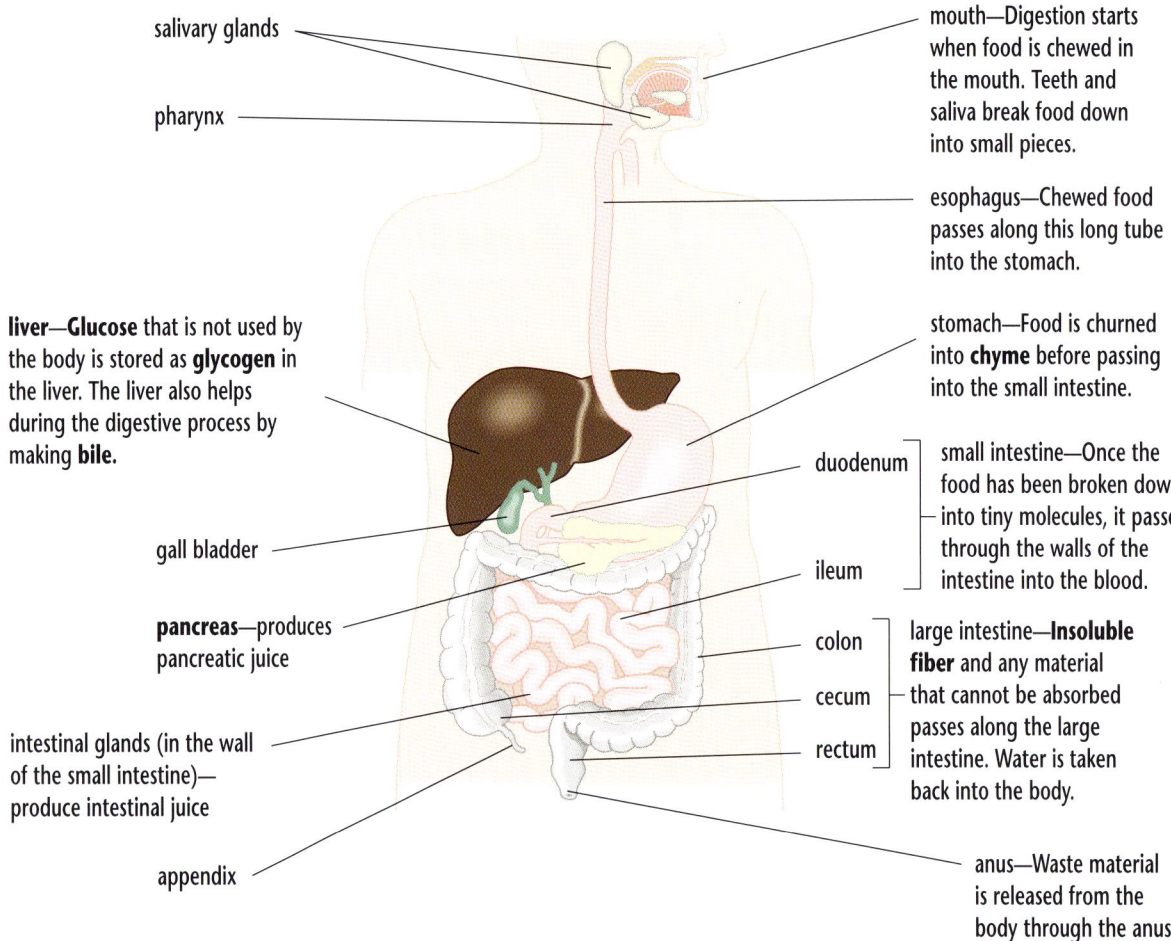

salivary glands

pharynx

liver—Glucose that is not used by the body is stored as **glycogen** in the liver. The liver also helps during the digestive process by making **bile.**

gall bladder

pancreas—produces pancreatic juice

intestinal glands (in the wall of the small intestine)—produce intestinal juice

appendix

mouth—Digestion starts when food is chewed in the mouth. Teeth and saliva break food down into small pieces.

esophagus—Chewed food passes along this long tube into the stomach.

stomach—Food is churned into **chyme** before passing into the small intestine.

duodenum

ileum

small intestine—Once the food has been broken down into tiny molecules, it passes through the walls of the intestine into the blood.

colon

cecum

rectum

large intestine—**Insoluble fiber** and any material that cannot be absorbed passes along the large intestine. Water is taken back into the body.

anus—Waste material is released from the body through the anus.

Fats have to be broken down so that they are small enough to be absorbed. Fats are **insoluble** in water, so a substance called **bile** does the job of breaking up fats and oily foods. The fats can then **dissolve** in water. Bile is made by the liver and stored in the gall bladder. After a meal, bile is released and passes into the small intestine where it begins breaking up fats that have just been eaten. This process is called **emulsification,** and bile mixes with the fat in the food, breaking the large fat droplets into smaller droplets. This makes it easier for the body to absorb it.

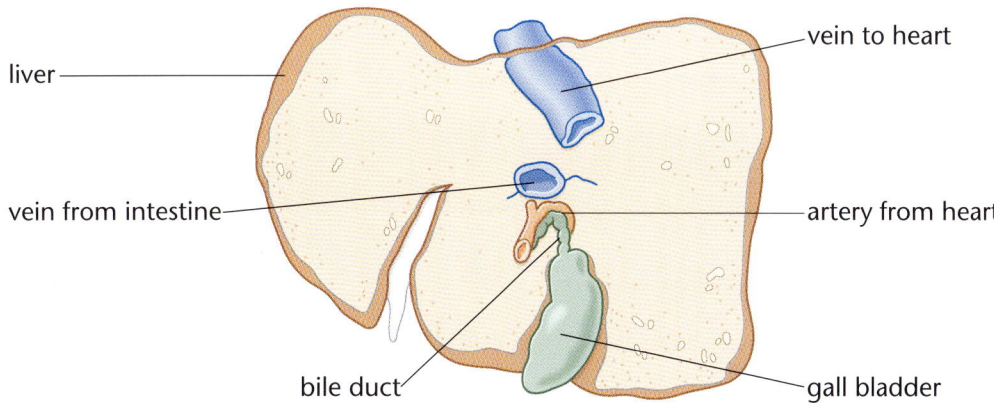

liver

vein to heart

vein from intestine

artery from heart

bile duct

gall bladder

This cross section of the liver shows what the liver looks like on the inside.

Fat Storage Areas

Your body can absorb and store fat in its liver, muscle, and fat cells. Fat cells make up fat tissue, which is found under the skin and cushioning organs such as the liver and kidneys. In liver and muscle cells, fat is stored as microscopic droplets of fat. You can see these tiny droplets of fat in an animal's muscles as white streaks on meat.

How Does the Body Absorb Fats?

The way the body absorbs fats is through a series of small **chemical reactions**. **Enzymes** are substances in the body that speed up a chemical reaction. They work together with **hormones** and **proteins** to make fats that you have eaten usable for energy.

Once **bile** has broken down fats in your small intestine, your body needs to absorb them. Your **pancreas** sends out enzymes that attack the fat. The enzymes break fat down into **fatty acids** and **glycerol.** These are substances that can be absorbed into the cells that line your intestines.

Inside the **cells** of the small intestine, the fatty acids and glycerol are rebuilt into bundles called **triglycerides.** They have a **protein** coating that makes the bundles **dissolve** more easily in water. Triglycerides then travel around your body and pass into your veins and bloodstream.

Your body now needs to move the fats being carried in your bloodstream into cells, where it can use them or store them as **energy.** The triglyceride bundles can be absorbed into fat, muscle, or liver cells. Once again, the body releases enzymes to break them down into fatty acids again. These enzymes are triggered by a **hormone** known as **insulin.**

Your Body Needs Fat for Energy

Your body needs to move fat to different areas, either to be used as energy or to be stored for later use. The fats that you eat need to be broken down because only very small particles can move between cells across the cell wall.

What is insulin?

Insulin is a hormone made in the body by the pancreas. A hormone is a substance that tells your body to do something. Insulin tells your body how to store energy. When you eat a meal or sweet snack, your body senses fatty acids, **amino acids**, and **glucose** in your intestine. Your brain tells your pancreas to make insulin. Insulin then encourages your **liver**, muscle, and fat cells to absorb the fatty acids, amino acids, and glucose.

If a person has too much insulin in the body, the enzymes that break down the fats are very active. The insulin levels tell them to work. If levels of insulin are low, the enzymes are not active. They are not being given the message to break down the fats.

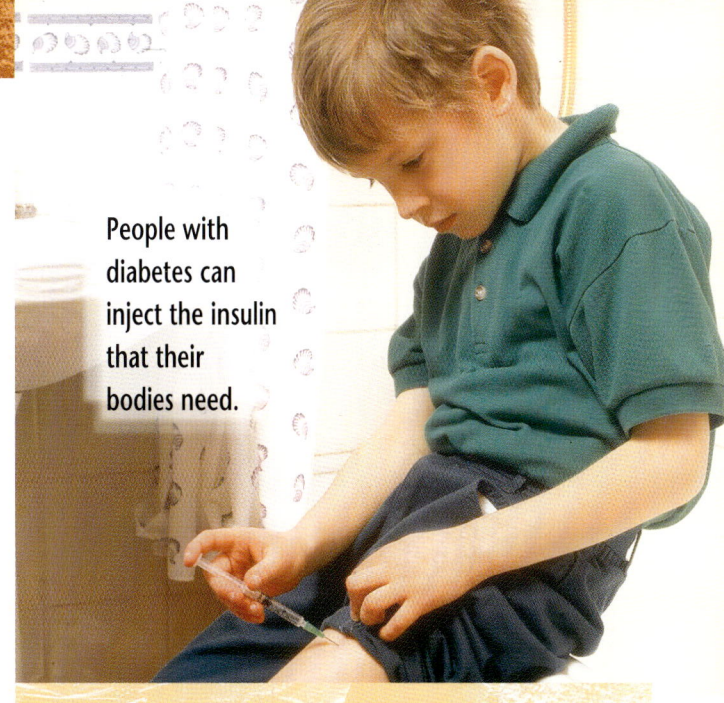

People with diabetes can inject the insulin that their bodies need.

Diabetes

Your body makes insulin when you eat. Normally, it can make all the insulin you need, but sometimes the body may not be making enough insulin. This is called diabetes, or **diabetes mellitus.** People who are diabetic may need to inject insulin.

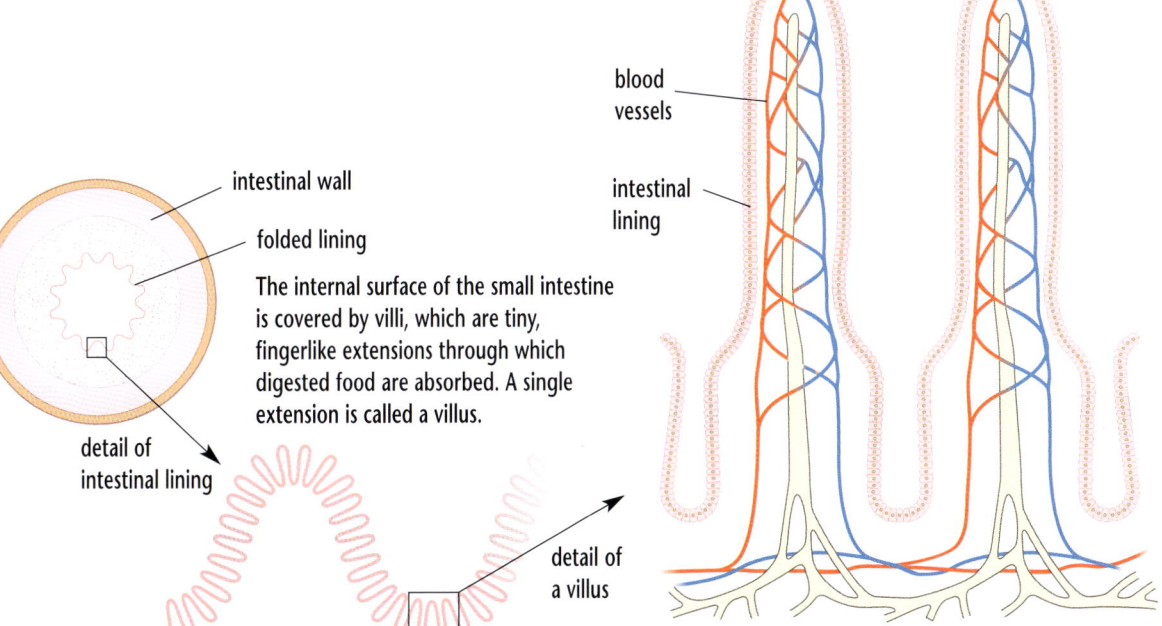

intestinal wall

folded lining

The internal surface of the small intestine is covered by villi, which are tiny, fingerlike extensions through which digested food are absorbed. A single extension is called a villus.

detail of intestinal lining

blood vessels

intestinal lining

detail of a villus

How Does the Body Store Fat?

Fat is stored in **tissue** that is made up of fat **cells.** Fat cells are like tiny little bags of fat. There are two types of fat cells: white fat and brown fat. White fat cells are large cells with one large fat droplet in them. Brown fat cells are smaller and contain several smaller fat droplets.

White fat is important for keeping your body warm and giving it **energy** and protective cushioning. Brown fat is found mainly in newborn babies and is important for keeping them warm.

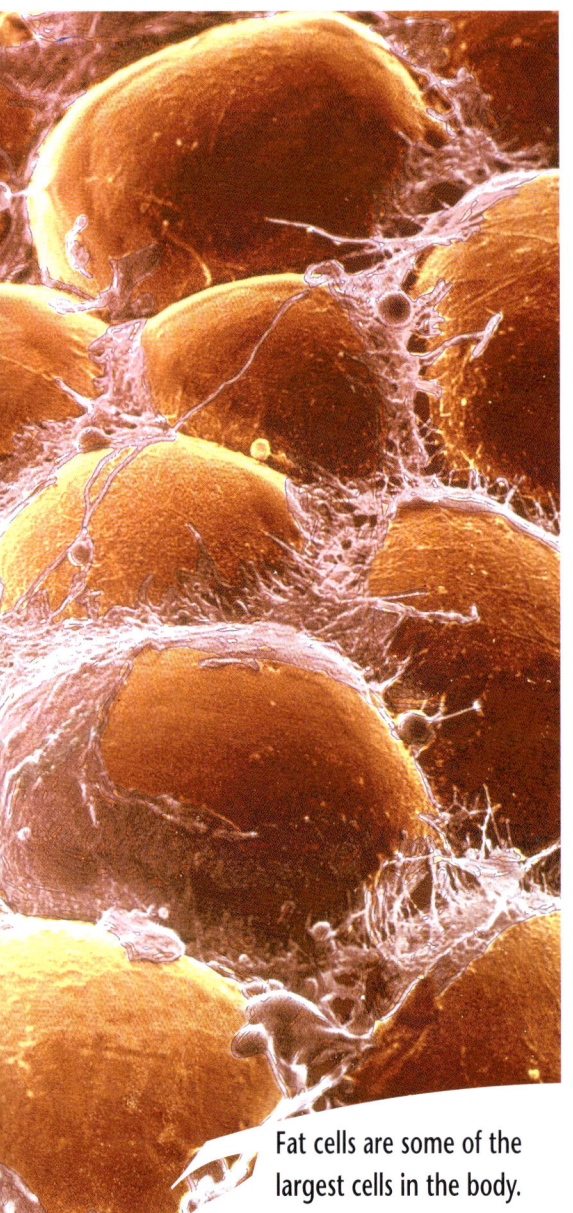

Fat cells are some of the largest cells in the body.

Babies and brown fat

Fat cells form in an unborn baby in the three months before it is born. Newborn babies do not have much white fat, so their bodies make warmth by using brown fat cells. Once babies start to eat more and grow, their bodies begin to store white fat in their white fat cells. It replaces the brown fat. Adults have few or no brown fat cells. After **puberty,** the body makes no more fat cells. As it stores more fat, the number of fat cells stays the same but the cells get bigger.

Some body fat is stored under your skin. This fat helps keep you warm because it does not let heat pass through it very easily. Fat also cushions and protects **organs** such as your kidneys.

Your body's fat cells are not very active. They do not **metabolize** fat, or use it to gain energy or build new body tissues. They just store it. When you eat a meal containing fats, your fat cells pick up spare fat in your bloodstream and store it.

Per ounce or gram, fats provide the most energy of all the nutrients. If we consume more energy than we use up, the excess energy is stored as fat. This is why eating too many fatty foods, which provide lots of energy, may lead to weight gain.

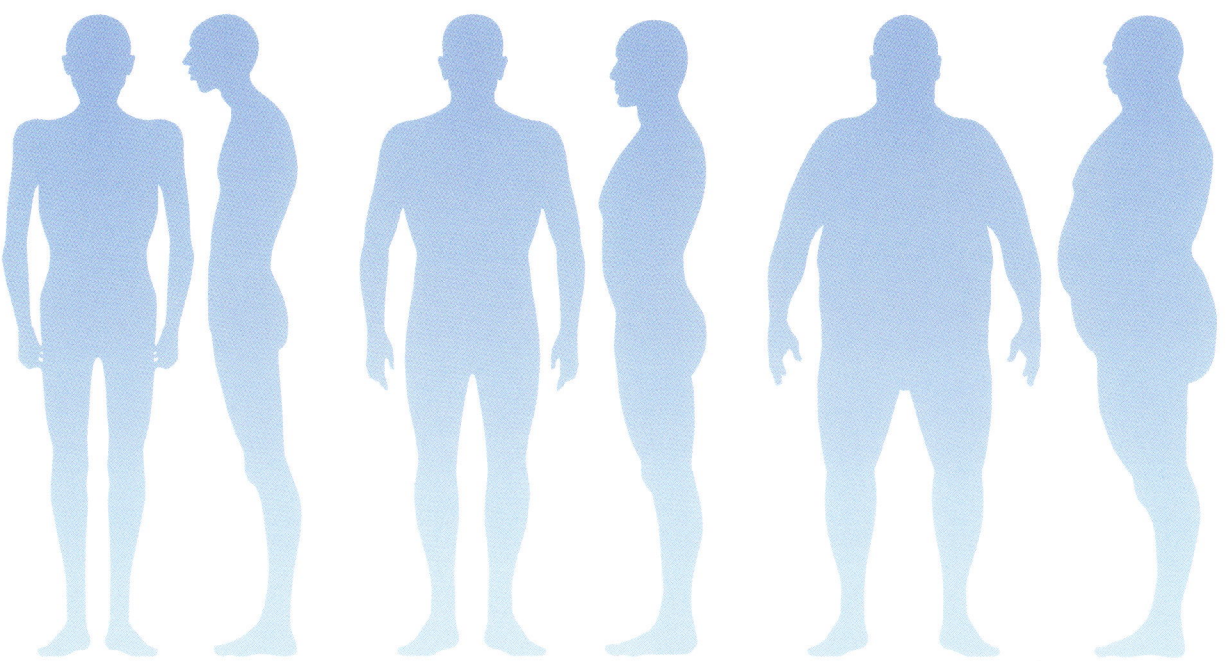

There are three main body types. From left to right they are: ectomorph (thin), mesomorph (muscular), and endomorph (fat).

Body Mass Index

Body Mass Index, or BMI, provides a guide based on height and weight that determines if a child is overweight or underweight. It is measured by dividing the child's weight by the child's height squared, or weight $\div$ height^2. The weight must be measured in kilograms and the height should be measured in meters for the calculation to be correct. The BMI is then located on a chart that compares the average BMIs of other children of the same age. The BMI will then let parents and doctors know if a child's weight is appropriate for his or her height.

How Does the Body Get Energy from Fats?

Your body gets **energy** from the food you eat. It can get energy from foods containing fats, **carbohydrates,** and **proteins.** For a healthful diet, you need to get energy from different food sources. Fats are the richest source of energy. They provide more than twice as much energy per gram as carbohydrates or proteins do. We measure the energy we get from food in **calories.**

Compare the Energy

- 1 gram of fat provides 9 calories of energy.
- 1 gram of carbohydrate provides 4 calories of energy.
- 1 gram of protein provides 4 calories of energy.
  (0.035 oz = 1 g)

When we are active, our bodies burn fat like cars burn fuel.

Why do you need energy?

Your body needs energy just to keep warm. You need energy for basic body processes such as breathing, digesting food, and sleeping. Young people need extra energy for growing. Your body is using energy even when you are resting. It uses energy to grow new cells and repair damaged cells.

The amount of energy you are using each minute when you are resting is called your **basal metabolic rate** (**BMR**). An adult uses about one calorie of energy for basic body processes each minute. Men usually have higher BMRs than women because they are usually more muscular than women. Older people tend to have lower BMRs because they have lost some of their muscle as they have aged. Babies and young children have a high BMR because they are using energy to grow. Three quarters of the energy we use is taken by the BMR when we are just resting. The rest of our energy needs depends on our body weight and how active we are.

You need to get enough energy from the food you eat to stay a healthy weight for your size and give you energy for activities. You could try to live on candy bars, but it would not be a good idea. They would give you energy, but they would not provide the other **nutrients** you need. In **developed countries,** such as the United States, fats provide about 40 percent of the energy we get from our food. But health experts in the United States recommend that fats should provide no more than 30 percent of our total energy intake. We should get about 55 percent of our energy from carbohydrates and the rest from proteins.

How Much Do You Need?

The energy you need depends on your age and how active you are. A baby boy just under a year old needs about 920 calories of energy a day. A baby girl needs about 860 calories of energy a day. An eleven-year-old boy who is 4 feet 10 inches (1.47 meters) tall and weighs 85 pounds (38 kilograms) needs a minimum of 1,261 calories per day. A girl of the same age and weight needs at least 1,225 calories per day.

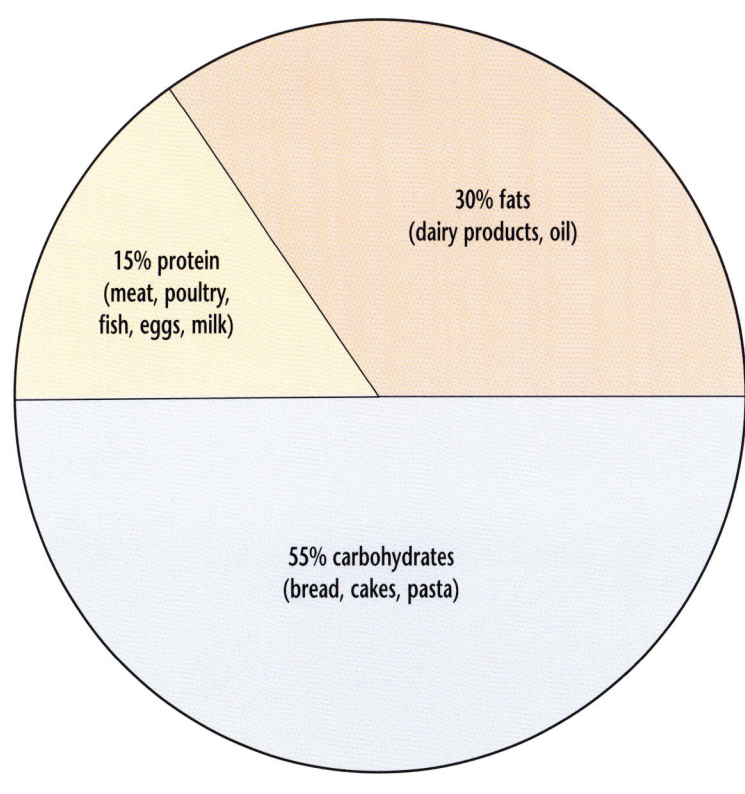

30% fats
(dairy products, oil)

15% protein
(meat, poultry,
fish, eggs, milk)

55% carbohydrates
(bread, cakes, pasta)

The chart shows the percentages of total energy intake that should come from fats, carbohydrates, and protein.

How Does the Body Turn Fats into Energy?

The more active you are, the more **energy** your body needs. Adults use about one **calorie** each minute when they are just resting. Between meals, the body can use **fatty acids** from your bloodstream as well as **glucose** to give you energy. When you start to become active, you will need more energy. Walking quickly, you will use between two to four calories each minute. Playing a sport or running, you will use about seven to nine calories a minute.

Where does the energy come from?

As you become more active, your body begins to take energy from **glucose** (sugar) in your blood. Glucose is the body's main source of energy. Some **cells** in your body, like your brain cells, can only get energy from glucose. Glucose comes from **carbohydrate** foods that you eat. It is stored in your **liver** as **glycogen.**

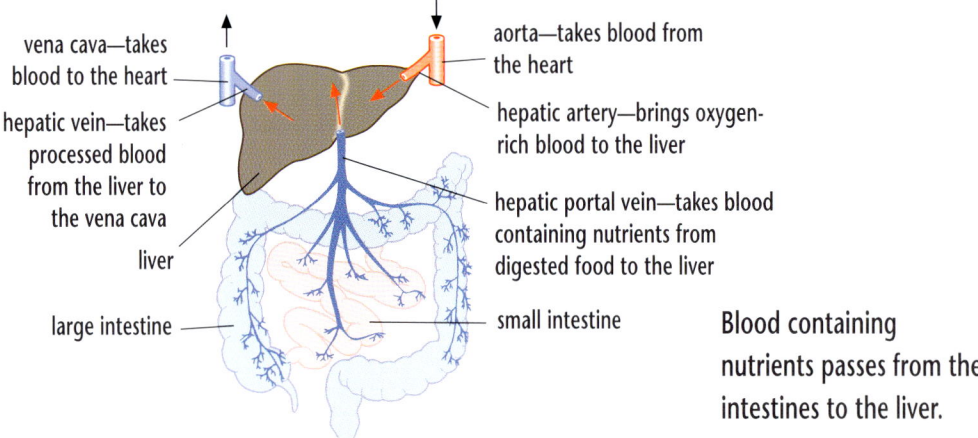

vena cava—takes blood to the heart

hepatic vein—takes processed blood from the liver to the vena cava

liver

large intestine

aorta—takes blood from the heart

hepatic artery—brings oxygen-rich blood to the liver

hepatic portal vein—takes blood containing nutrients from digested food to the liver

small intestine

Blood containing nutrients passes from the intestines to the liver.

The **liver** acts like a chemical factory unlocking energy from your food. A vein carries blood containing food **nutrients** from the intestine to the liver. The liver filters the blood. It removes chemical waste and stores useful substances such as glucose and **vitamins.**

Energy for exercise

As you exercise, your body uses stored carbohydrates and fats to give you energy. It first takes energy from **glycogen** in the muscles, which it has stored from carbohydrate foods. The body can only store a certain amount of glycogen. The bigger your muscles are, the more glycogen they can hold.

When you are resting or doing such activities as walking or writing, your body mainly **metabolizes** fat as a source of energy for your muscles. When you start to exercise the working muscles need a more immediate source of energy so the body switches from using fat to stored glycogen as the source of energy. Once all the glycogen has been used up, you start to feel tired and are forced to slow down as the muscles can only use fat for energy.

A trained athlete's body burns fat more quickly than a nonathlete's body. The longer and harder athletes exercise, the more fat is burned for energy. After an hour's hard exercise, up to 75 percent of their energy may be provided by fats.

How Much Energy?

An adult uses an average of 2,390 calories of energy each day, depending on how active he or she is. A young man will use about one and a half calories of energy each minute just sitting down. Walking slowly, he will use about three calories of energy each minute. If he plays soccer, the energy he uses will increase to seven calories each minute.

Peanuts and Fat

Your body metabolizes fat to give it energy. When you eat a peanut, your body digests the fat in the peanut, which will be stored or used as energy.

Athletes who compete all over the world use a lot of energy every day.

Fats for Vitamins

You need some fats in your diet because they provide you with the **vitamins** A, D, E, and K that you need to stay healthy. All these vitamins are **fat soluble,** which means they are **dissolved** in the fats in your food. They are found in foods such as meat, dairy products, egg yolks, vegetable seed oils, and leafy green vegetables. Some foods, such as milk and margarine, have vitamins A and D added to them. If you ate a very low-fat diet for a long time, your body might not get enough of these important vitamins.

When you digest foods containing fats, fat-soluble vitamins are carried into the intestine. There, your body can absorb them.

Vitamin A

Vitamin A is found in oily fish, fish-liver oils, eggs, dairy products, and vegetables, such as carrots and spinach. Your body needs it for growth, healthy skin, and good eyesight. A shortage of vitamin A can result in problems with eyesight and even blindness. Your body can only absorb vitamin A by getting it from fats in foods that you eat. You can get vitamin A from animal or vegetable foods. Animal sources of vitamin A are six times stronger than vegetable sources and can be toxic if you have too much.

A Lack of Vitamin A in Developing Countries

The World Health Organization believes that between 100 and 140 million children in **developing countries** suffer from a lack of vitamin A. It is a problem in 118 countries, especially those in Africa and Southeast Asia. Not getting enough vitamin A can cause blindness and result in death and disease from infections. The World Health Organization runs programs that give liquid vitamin A to children while they are being **immunized** against diseases such as polio.

The sunshine vitamin

Vitamin D is sometimes called the sunshine vitamin because sunlight is one source of it. It can enter your body through your skin. It is also found in liver, oily fish, fish-liver oils, margarine, and some breakfast cereals. Vitamin D is important for forming healthy bones. A lack of vitamin D can lead to bone problems such as rickets. Rickets is a disease in which bones do not grow well.

Vitamin E

You need vitamin E for healthy skin. It is also important for long-term health. It is found in eggs, butter, milk, nuts, vegetable oils, and seeds.

Vitamin K

Vitamin K is important for healthy blood and blood flow. It is found in green vegetables, tomatoes, eggs, and some cereals.

Stored vitamins

Your body can store fat-soluble vitamins in your liver and in your fat cells until it needs them. You can make sure you have enough vitamins stored by eating more of the right kinds of food. Some groups of people such as pregnant women, growing children, and senior citizens may need extra vitamins. But it is important not to get too many fat-soluble vitamins, because they can build up in the liver and have toxic effects.

These foods are part of a healthy diet because they contain important vitamins.

Essential Fatty Acids

Your body can transform **fatty acids** from foods into different types of fatty acids to be used in different **cells** as it needs them. But there are two kinds of **polyunsaturated** fatty acids (PUFAs) that your body cannot make, but which are essential for your health. These are called **essential fatty acids (EFAs)** and you can only get them by eating foods that contain them. One EFA is called linoleic acid. It is found in vegetable seed oils such as sunflower oil, soybean oil, and in small amounts in animal fats. The second EFA is linolenic acid, and it is found in small amounts in vegetable oils.

Polyunsaturated fatty acids

Scientists group PUFAs into two families, called **omega-3 fatty acids** and **omega-6 fatty acids**.

Foods such as oily fish, nuts, and leafy green vegetables are all sources of essential fatty acids.

22

Polyunsaturated fatty acids are found in certain types of fish, green leafy vegetables, seeds and nuts, beans, and grains. The omega-3 fatty acids are found in soybeans, walnuts, linseed and flax oil, dark green leafy vegetables, and fish such as salmon, tuna, and mackerel. The omega-6 fatty acids are found in vegetable fats such as sunflower oil and margarine. Linoleic acid belongs to the omega-6 family, and linolenic acid belongs to omega-3.

Why do you need EFAs?

Many types of body **cells**, including your brain, nerve, skin, and hair cells, need EFAs to keep them healthy. Your brain could not work without EFAs. They also help your body make chemical substances needed for many body processes, such as helping blood clot and form a scab when you have hurt yourself.

EFAs help your body's **immune system** fight **bacteria** and **viruses**. Your body uses EFAs to build cell walls and keep them healthy. Strong, healthy cell walls help keep bacteria and viruses from getting inside cells and causing disease.

Omega-3 fatty acids may help protect you against heart disease by keeping your blood flowing smoothly. They may also keep your joints healthy and protect you against diseases that affect the joints, such as arthritis.

Storing Fatty Acids

Fatty acids react easily with chemicals in the body, which is why they have health-giving powers. But this also means they can make the foods in which they are found change chemically. Nuts and seeds containing EFAs can spoil when the fats in them react with oxygen in the air. They need to be stored in an airtight container and kept in a cool place.

Fats in Processed Foods

Foods such as seeds and grains contain natural **fatty acids.** When we eat them, we also get calcium, which the body needs to help keep the bones healthy. These foods also provide **antioxidant vitamins** that the body needs to keep **cells** healthy.

However, some fatty acids change when food is **processed.** Fats are used in many processed foods such as cookies and frozen dinners. Manufacturers process natural foods to get them to look and taste the way they want. They heat the food and use chemicals called additives to change the color, texture, taste, and smell of the food.

Hydrogenated fats

One of these processes is called **hydrogenation.** It is used to turn **unsaturated** fats, which are liquid at room temperature, into a solid form. Margarine is an example of a fat that has gone through hydrogenation. Hydrogenation involves adding **hydrogen** into fatty acid chains so they become **"saturated"** with hydrogen.

Hydrogenation changes some of the unsaturated fats into saturated fats.

The unsaturated fats that do not become saturated become **trans fats** instead. Trans fats keep the different types of fats together so they stay solid at room temperature. Trans fats also make fats less likely to spoil, so they will last longer in the refrigerator or on a cupboard shelf.

Pastries are often made with hydrogenated fat.

On food labels, hydrogenated fats are often listed as "hydrogenated vegetable oil" or "partially hydrogenated vegetable oil." Hydrogenated fats are often used to deep fry foods such as french fries and onion rings. They are also used in snack foods.

Trans fats

Natural trans fats are sometimes found in small amounts in foods such as meat and dairy products. But most trans fats are found in fried foods and processed foods that are high in fat, such as cookies, cakes, crackers, and margarine.

A high intake of trans fats in the diet increases levels of **cholesterol** in the blood. This increases the risk of heart disease. Most manufacturers have removed trans fats from their products.

Frying Fat
Fats can change chemically when they are heated at high temperatures. Frying can make some cooking oils react with oxygen and form chemicals called **free radicals.** These can damage body cells and cause them to age. Butter and olive oil are less likely to form free radicals at high temperatures.

Fats can change chemically when they are heated at high temperatures.

Fat and Cholesterol

Cholesterol is a type of fat found in your body. Your body needs some cholesterol to build **cell** walls and brain and nerve **tissue.** It also uses cholesterol to make **hormones** needed for basic processes like digestion. Your body can make about 75 percent of the cholesterol it needs from the dietary fat that you eat. One of the things cholesterol is used for is to be converted into **bile,** which helps you digest and absorb fat from the diet.

Dietary cholesterol

You also take in some cholesterol from the things you eat every day, or your diet. Dietary cholesterol is found in animal foods, such as egg yolks, meat, liver, some shellfish, and milk. Unless you have a very high intake of cholesterol, dietary cholesterol has little effect on cholesterol levels in your blood. Blood cholesterol levels are mainly affected by intake of **saturated** fats.

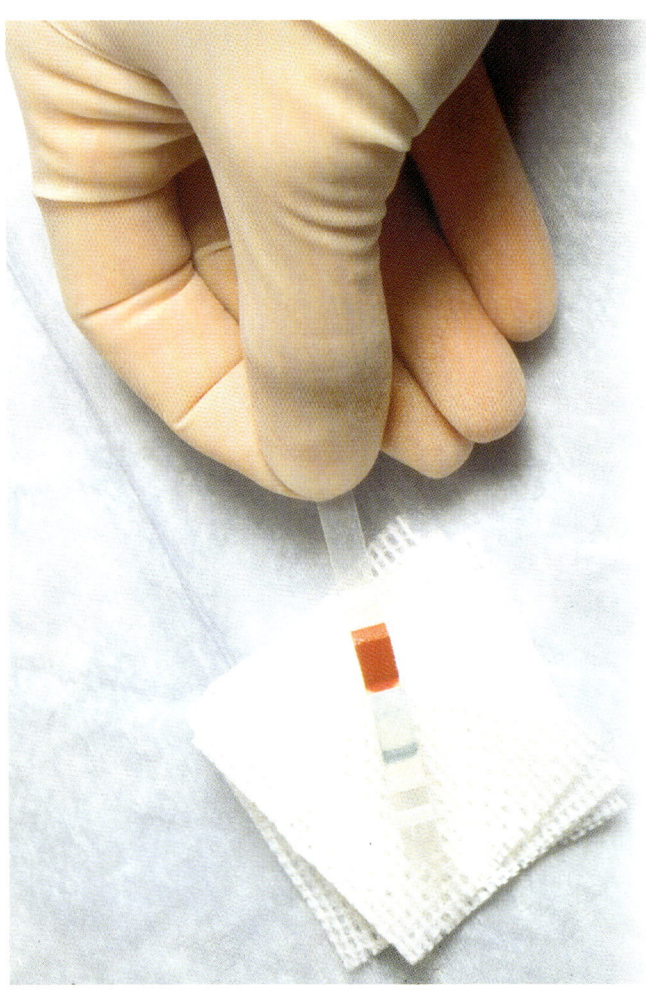

Blood cholesterol levels

"Good" cholesterol (called **HDL cholesterol**) helps carry other cholesterol and fats away from the **arteries** to the liver where they can be broken down. HDL cholesterol can help protect the body against heart disease. But "bad" cholesterol (called **LDL cholesterol**) can slowly build up in the walls of the arteries that feed blood to the heart and brain. It can form fatty deposits that clog and narrow the arteries. If a **blood clot** forms in the arteries, it can block the flow of blood to the heart and cause a heart attack. It if blocks the flow of blood to the brain, it can cause a **stroke.**

A blood cholesterol test shows the level of cholesterol present in the blood.

Controlling cholesterol

Doctors measure the amount of cholesterol in the blood in milligrams per deciliter (1 milligram = 0.00035 ounce; 1 deciliter = 3 ounces). For adults, doctors recommend that the amount of cholesterol in the blood be kept below 200 milligrams per deciliter. In the United States, many people eat foods that are high in animal fats, which can often lead to high blood cholesterol. This increases the risk of heart disease, so doctors recommend lowering the amount of cholesterol in the blood by changing eating habits and lifestyle.

Lowering cholesterol

Some foods, such as garlic, can help lower blood cholesterol. Others, like flaxseed, can lower the amount of "bad" LDL cholesterol and increase the "good" HDL cholesterol in the blood. Certain margarine spreads can help lower blood cholesterol. They have a plant ingredient added to them called plant stanol ester, which keeps the body from absorbing cholesterol when we eat. In some cases, doctors may also recommend special drugs to lower high blood cholesterol.

Low-fat foods can help control the amount of dietary cholesterol.

Adults Can Control Cholesterol by:
- losing extra weight
- getting more exercise
- stopping smoking
- drinking less alcohol
- avoiding foods high in saturated fats
- eating a low-fat diet
- eating foods that contain fiber, such as fruits, vegetables, and whole grains

Fats and Overeating

When you take in more **energy** than you use up, the excess energy is stored as fat. This can lead you to become overweight. Foods containing a lot of fat provide more energy than those that contain mainly either **carbohydrate** or **protein.** People like high-fat foods such as burgers and chips, cakes, and doughnuts because their fat content gives them a good taste. This is what can tempt some people to overeat.

There are many other reasons that people may become overweight. A lack of physical activity is one of the main reasons why kids become overweight or **obese.** The recommended time for a child to have moderate physical activity is about 60 minutes a day for most of the days of the week. Some recent studies partly blame the **genes** that a child gets from his parents as a possible cause for obesity.

Some ways to avoid overeating and obesity include eating meals as a family more often and at a specific time daily, avoiding using food to reward good behaviors or actions, and eating your meals slowly. You should also try to eat only healthful snacks, drink water instead of soft drinks, eat at least five servings of fruits and vegetables every day, and have a healthful breakfast every day.

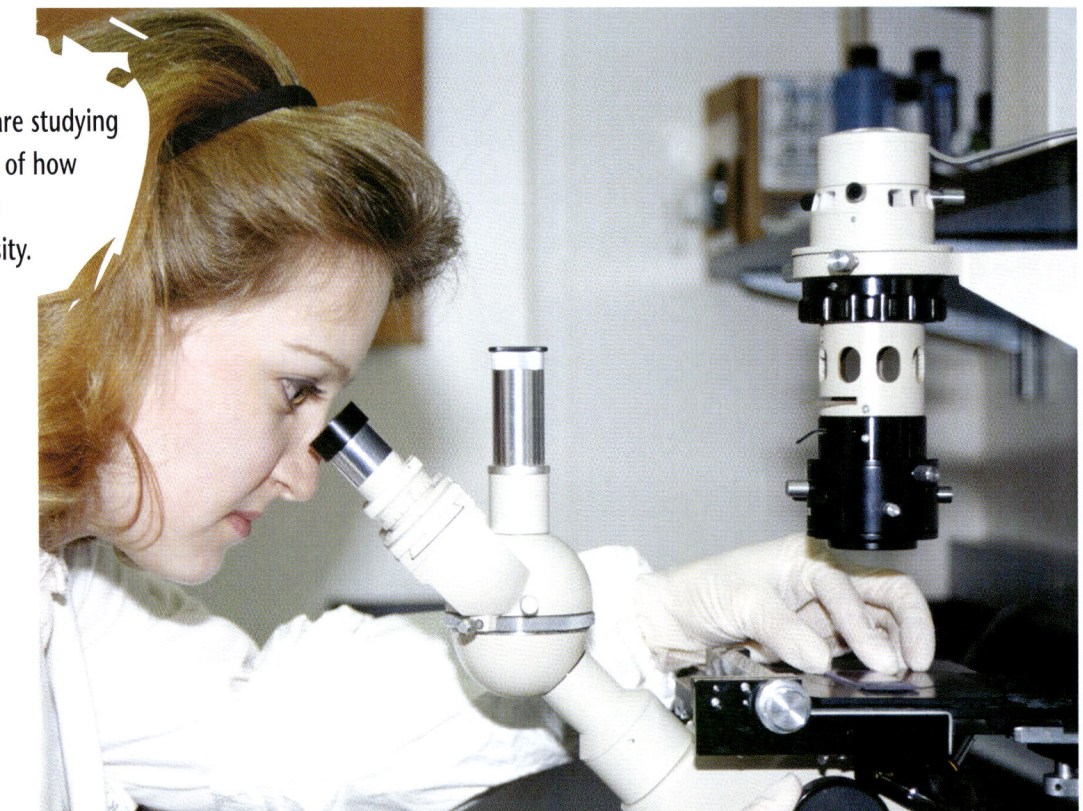

Scientists are studying the effects of how genes may cause obesity.

Fat and the basal metabolic rate

Fat **cells** are not as active as other body cells, such as muscle cells. They burn a lot less energy than muscle cells. So, the more fat and less muscle a body carries, the lower the **basal metabolic rate** of that person will be.

When a person wants to lose weight, he or she should reduce energy intake and become more physically active. That way, the person is using up more energy than he or she is taking in. People wanting to lose weight should not reduce their energy intake by too much or they will start to lose muscle as well as fat. It is important that people who are dieting get a balanced intake of all the essential **nutrients**. Therefore, the diet should be high in fruits and vegetables and include high-**fiber**, low-fat foods.

Obesity is a growing problem in many parts of the **developed world.**

There are several problems that have developed recently as a result of the increasing rate of obese and overweight children. There has been an increase in children who have heightened risk factors for heart disease as a result of obesity and being overweight. These risk factors include high cholesterol and high blood pressure. However, the most immediate consequence of being overweight is that children who are overweight are often made fun of by other children. This leads to low self-esteem and sometimes depression for the children that are being made fun of.

Fats and Obesity

We need to eat just enough fat to keep ourselves healthy and give us **energy.** If we get lots of exercise, our body will use the fats we eat to give it energy. By balancing the number of **calories** we eat and the amount of energy we use, we will stay at a healthy weight.

Fat Rats!

In a study done at the University of Illinois, one group of rats was fed a diet of 42 percent fats. The other group ate a low-fat diet. Both groups were allowed to eat as much as they wanted. After 60 weeks, the low-fat group was still lean and sleek. Rats in the high-fat group were overweight and had up to 51 percent body fat.

Body mass index

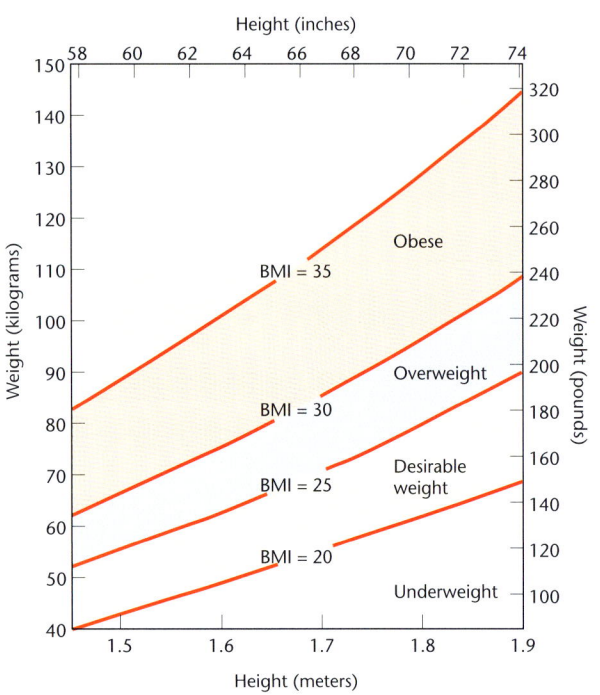

Are you obese?

Work out your BMI from the chart or use the formula

BMI = Weight in kilograms ÷ (height in meters)^2

A BMI of less than 20 is underweight
20–24.9 is desirable
25–29.9 is overweight
30+ is obese

Fat lifestyles

It is normal for children and teenagers to store fat because they are still growing. But because of the diet and lifestyle we have today, many children and adults are overweight. Popular foods like burgers and fries, pizza, and ice cream are high in fat. Many **processed** foods have more fat in them than you might think. Also, people get much less exercise than they used to. In your great-grandparents' day, many people used to walk to work and school. Now many people go by car, bus, or train. In the past, people were more likely to spend their spare time doing something active. Now they spend more time watching television or using computers. Sitting in front of a TV or computer screen uses about as much energy as sleeping!

Obesity

More people are becoming **obese.** In the United States, nearly one-third of adults are obese. Among children and teens ages six to nineteen, fifteen percent are overweight. Children who are overweight are likely to be overweight or obese when they grow up. The main causes of obesity are a poor diet and a lack of exercise.

Scientists are researching drugs to fight obesity, but the best way of tackling it is to cut the amount of calories eaten each day, eat a low-fat diet, and get more exercise.

Keep Fit Not Fat
Regular exercise helps burn off stored fat. It may also speed up the rate at which the body turns food into energy by as much as ten times.

Exercise keeps your body healthy, and it can be lots of fun.

Fats and Health Problems

Fats should make up no more than 30 to 35 percent of our diet, according to health experts. This will give us plenty of **energy** and all our **essential fatty acids.** But in the diet in **developed countries** such as the United States, where people eat a lot of meat and **processed** foods, about 40 percent of total **calories** come from fats.

Health problems

A high-fat diet can cause many health problems. Some of these are linked to **obesity.** Obesity increases the risk of heart disease, **stroke,** and some cancers. It also increases the risk of **diabetes mellitus,** in which the body is not able to use sugars normally. Doctors believe that an "apple" body shape, which carries most of its fat around the middle, is more likely to develop health problems than a "pear" shape, which carries most weight of its weight around the hips and thighs.

Obesity puts extra stress on many parts of the body, such as the bones, blood circulation system, and nerves. It increases the risk of osteoarthritis, a disease in which the joints become painful and swollen. Obesity can also lead to high blood pressure and heart and breathing problems.

Saturated fats are found in:
- **meat and dairy products**
- **some plant oils, like coconut and palm oil**
- **processed foods like cookies, crackers, and ice cream**

Fries are tasty, but if you eat them too often you will run the risk of damaging your health.

Cancer

Diets that are high in fat, especially **saturated** fats, have been directly linked with some cancers, including breast, colon, and skin cancers.

Doctors believe that about 30 to 40 percent of all cancers could be prevented by eating low-fat, high-**fiber** foods and getting regular exercise. To prevent and treat cancer, they recommend a diet that is low in fat and high in plant foods such as fresh fruits and vegetables.

Cancer Facts

Women in Japan have low rates of breast cancer compared to women in the United States. They eat a diet that is low in saturated fats and high in fish oils and soy **protein**. When Japanese women move to Western Europe or the United States and start eating more saturated fats, their rate of breast cancer increases.

Sushi—often made with raw fish—is a popular part of the Japanese diet.

Fats and Heart Disease

Eating too much **saturated** fat increases the risk of heart disease. Heart disease causes 40 percent of all deaths in the United States and Europe.

Blocking arteries

Having a high intake of saturated fats in the diet raises the level of **cholesterol** in the blood. The cholesterol can accumulate in the walls of the blood vessels in the heart. Once cholesterol levels start to build up, the cholesterol can cause the vessels to become narrow. This leads to a reduced flow of blood to the heart and may cause chest pain, known as angina. The chest pain happens particularly during exercise.

As the fats and cholesterol build up, the **artery** walls start to thicken and harden. This thickening and hardening of the artery walls is called atherosclerosis. The following habits and health problems can lead to atherosclerosis:

- eating too many saturated fats
- having high levels of blood cholesterol
- smoking
- not getting enough exercise
- being **obese**
- having high blood pressure
- having **diabetes**

artery wall

fat deposits

Artery walls can become blocked by fat deposits.

A Heart-Healthy Diet

To avoid heart disease, the American Heart Association suggests staying away from foods that are high in saturated fat and cholesterol, such as steak, pork ribs, whole milk, and eggs. It recommends instead that people eat lower-fat meats such as chicken and turkey, low-fat or skim milk, and plenty of fruits, vegetables, and whole grains.

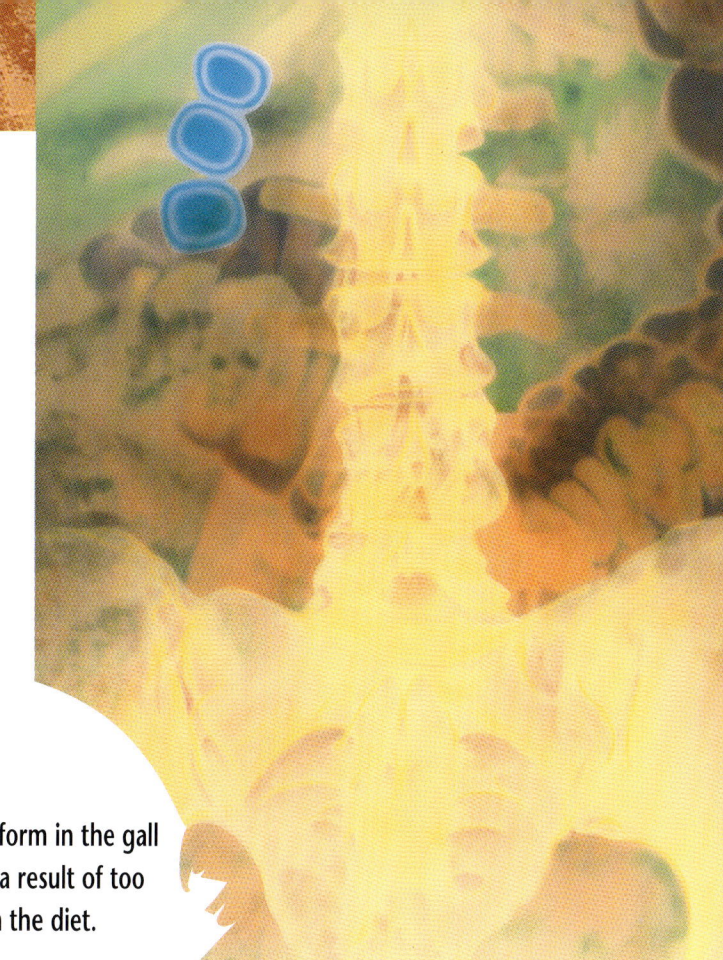

What Are Gallstones?

Your liver makes bile to help you break down fats in your food. But if you eat too many fatty foods, your liver can begin to make too much bile. Bile can then build up in your gall bladder. As it builds up, it can start to harden and form gallstones. These can cause pain and they may have to be removed surgically. Low-fat diets can help prevent gallstones from forming.

Gallstones form in the gall bladder as a result of too much fat in the diet.

Angina and heart attack

As the arteries narrow and the walls harden, it is harder for blood to flow around the body. The heart has to work harder to try and pump blood to the muscles when exercising. People who have narrowed arteries may get pains in the legs because the muscles are not getting enough blood and **oxygen.** They may also experience chest pains if not enough blood and oxygen are reaching the heart. This chest pain is called angina.

Eventually, the thick, sticky blood may form a **blood clot** that blocks an artery. If it blocks an artery that leads to the heart, it can cause a heart attack. If it blocks an artery leading to the brain, it can cause a **stroke.**

Recovering from a Heart Attack

People who suffer heart attacks must make several changes in their diets and lifestyle. First of all, a diet in which the intake of fats is strictly controlled is a must. This includes reducing the intake of high-fat meats, eggs, and other foods high in saturated fats and cholesterol. Also, people who smoke must stop immediately after a heart attack. People who start smoking again after a heart attack double their chances of suffering another heart attack.

Fats That Prevent Disease

A high intake of **saturated** fats and **trans fats** can raise blood **cholesterol** levels and lead to heart disease. But some **unsaturated** fats may help protect against heart disease by lowering "bad" cholesterol and raising or maintaining "good" cholesterol in the blood.

The good fats

Monounsaturated oils such as olive oil and **polyunsaturated** oils, such as soybean oil, may lower blood cholesterol levels. They also provide **vitamin** E, which may protect against heart disease. Nuts, especially walnuts and almonds, are also rich in polyunsaturated **fatty acids,** which can protect against heart disease and lower blood cholesterol. Monounsaturated oils may also increase levels of the "good" **HDL cholesterol** and help protect against heart disease.

Patients recovering from a heart attack are advised to eat a healthier diet.

Diets around the world

The diet typically eaten in the United States contains high levels of saturated fats from meat and dairy foods and also in baked and fried foods. In other parts of the world, including Asia and Africa, meat is used in much smaller amounts or is saved for special occasions. The diet in Asian and African countries is based on starchy **carbohydrates** such as rice and couscous. Low levels of saturated fats are linked with lower rates of heart disease.

The Mediterranean Diet
People living in Mediterranean countries such as Greece, Italy, and Spain are much less likely to develop heart disease than Americans. Mediterranean people eat mainly unsaturated fats. They only eat red meat a few times a month, and they eat only small amounts of fish, poultry, and eggs. Mediterranean people eat lots of starchy carbohydrates such as bread, pasta, and potatoes and plenty of fresh fruit and vegetables. They cook with olive oil, which is high in **monounsaturated** fats, and they eat nuts, seeds, and fish that are high in polyunsaturated fats.

Fish is good for the heart

Omega-3 fatty acids, which are found in fish oils, may help improve blood flow to the heart. Eating oily fish, such as mackerel or salmon, twice a week will provide about 0.035 ounce (1 gram) a day of omega-3 fatty acids. This may prevent **blood clots** from forming and protect against heart disease.

Who Has a Healthy Diet?

The Japanese diet contains just over 30 percent fat, compared with about 40 percent in the United States. Japanese cooking is based on rice, fresh vegetables, and oily fish—all of which are rich in polyunsaturated fatty acids. Japan has one of the lowest rates of heart disease in the world. The Inuit people of Greenland also have a low rate of heart disease. Their diet is based on fish and sea animals such as seals that are low in saturated fat and high in omega-3 fatty acids.

Asian diets are based on foods that are high in starchy carbohydrates and low in fat.

Getting Enough Fat

Eating too much fatty food is bad for your health, but not eating any fat would keep the body from getting important **nutrients.** We must eat some fat to get the **essential fatty acids** that the body cannot make itself. Shortages of these fatty acids can cause body growth problems, arthritis, and other health problems.

Essential Fatty Acids

It is important for your body to obtain essential fatty acids, or EFAs. EFAs are fats that humans cannot manufacture in the body, so we have to obtain them from food. EFAs are especially important for children, because they are so instrumental in proper growth and development. New information now also shows that EFA levels are significantly lower than average in children who are hyperactive. Scientists have said that hyperactive children may benefit from consuming more food or supplements with EFAs.

We need to consume about 0.14 ounces (4 grams) of **omega-6** fatty acids a day. This is about two teaspoons of sunflower oil or a few almonds or walnuts. We also need about 0.035 ounces to 0.071 ounces (1 to 2 grams) of **omega-3** fatty acids a day. This can be obtained from eating about 3.5 ounces (100 grams) of oily fish. Omega-3 fatty acids help to build healthy eye **tissue** and to keep blood flowing properly.

You can buy a variety of low-fat dairy products like these.

Fat for vitamins

The body needs at least 25 grams of fat a day to provide and absorb enough of the **fat-soluble vitamins** A, D, E, and K. When fats are removed from dairy products to make low-fat yogurt, skim milk, or cottage cheese, these foods lose much of their vitamin A. The body can make extra vitamin A from beta-carotene, which is found in leafy green vegetables, orange vegetables, and orange and

Lack of Vitamin A and Blindness

In Africa, many people do not have enough money to eat meat every day. If babies and young children do not get enough fats from plant foods, they can develop a vitamin A deficiency, or shortage, which can lead to blindness.

yellow fruits. But the body needs fats in the diet to be able to absorb beta-carotene and fat-soluble vitamins.

Fats and growth

Fats are needed to help babies grow and also during **puberty** when the body is growing and developing quickly. In its first year, a baby grows to three times what it weighed when it was born. Fat is a natural part of breast milk. It provides about half of the **calories** a baby gets each day until the baby is about one year old and eating solid foods.

Babies should not be fed cow's milk until after they are one-year-olds. They should be fed special formula milk or breast milk.

Babies eventually stop drinking milk and eat solid food such as prepared baby foods.

Balancing Fats in the Diet

A healthful diet needs to be low in **saturated** fats and have a good balance of **unsaturated** fats, including **omega-6** and **omega-3 fatty acids.** But in nations such as the United States, most of the fats in our diet are saturated fats from red meat, dairy products, and sweets.

Health recommendations

Health experts recommend that most Americans cut down on eating meats and dairy products that are high in saturated fats and eat more unsaturated fats from fish and plant foods. We also need to limit the amount of **cholesterol** in our diet, because saturated fats raise blood cholesterol levels, which can lead to heart disease.

We can get a healthful balance of fats in our diet by:
- using less butter
- drinking low-fat or skim milk
- eating no more than three to four eggs a week
- eating less fatty meat and more oily fish such as salmon
- cutting the fat off red meat and the skin off chicken
- avoiding fried foods
- using olive oil or another vegetable oil for cooking rather than hard fats such as butter or margarine
- avoiding **processed** foods that contain saturated or **hydrogenated** fats

Fats on food labels

Read food labels to find out how much fat is in food. Food labels list the number of grams of fat in a serving and may give a breakdown of how much of the fat is saturated, **polyunsaturated,** or **monounsaturated.** It also tells you what percentage of the daily recommended amount of fat is in each serving. Look for low levels of saturated fats and higher levels of unsaturated fats, especially the omega-3 polyunsaturated.

Cut down on fat by choosing low-fat foods. Compare the following to see what a difference it can make:

3.5 ounces (100 grams) of french fries have 0.24 ounces (6.7 grams) of fat

BUT   3.5 ounces of boiled potatoes have only 0.007 ounces (0.2 grams) of fat

3.5 ounces of roast beef have 0.74 ounces (21 grams) of fat

BUT   3.5 ounces of roast chicken (without skin) have 0.14 ounces (4 grams) of fat

Processed food

Seventy percent of the fat we eat is "hidden" in **processed** foods. Try to avoid eating too many processed foods, such as cookies, crackers, candy bars, and chips.

We can try to balance the high-fat foods we eat with low-fat foods to keep the total amount of fat we eat to no more than 30 percent each day. Look for low-fat products such as skim milk, low-fat or fat-free yogurt, reduced-fat peanut butter, and reduced-fat cheese.

Lowering cholesterol levels

According to the U.S. Food and Drug Administration (FDA), studies have shown that a plant substance known as plant stanol ester keeps the body from absorbing cholesterol. This can help lower "bad" LDL cholesterol levels. Spreads containing plant stanol esters can be used in place of margarine or butter. Another way to lower cholesterol is to cut down on eating foods high in saturated fats and other foods such as egg yolks that are known to raise blood cholesterol levels.

Stir frying is a healthy way of cooking because it only uses small amounts of oil.

Cooking with Fats

Choose healthy ways of cooking, such as grilling, steaming, and baking, rather than frying and roasting. Stir-frying uses less fat than deep-frying. Hard fats such as lard and butter are high in saturated fats, so it is healthier to cook with vegetables oils such as olive oil, canola oil, or corn oil. Never reuse old cooking oil or eat burned foods. They can contain harmful **free radicals,** which can damage your health.

Fats and the Food Guide Pyramid

The U.S. government recommends a balanced diet based on six main food groups. The Food Guide Pyramid below shows each of the groups we should eat from every day. Fats, oils, and sweets are also shown on the pyramid, but you eat them only in small amounts.

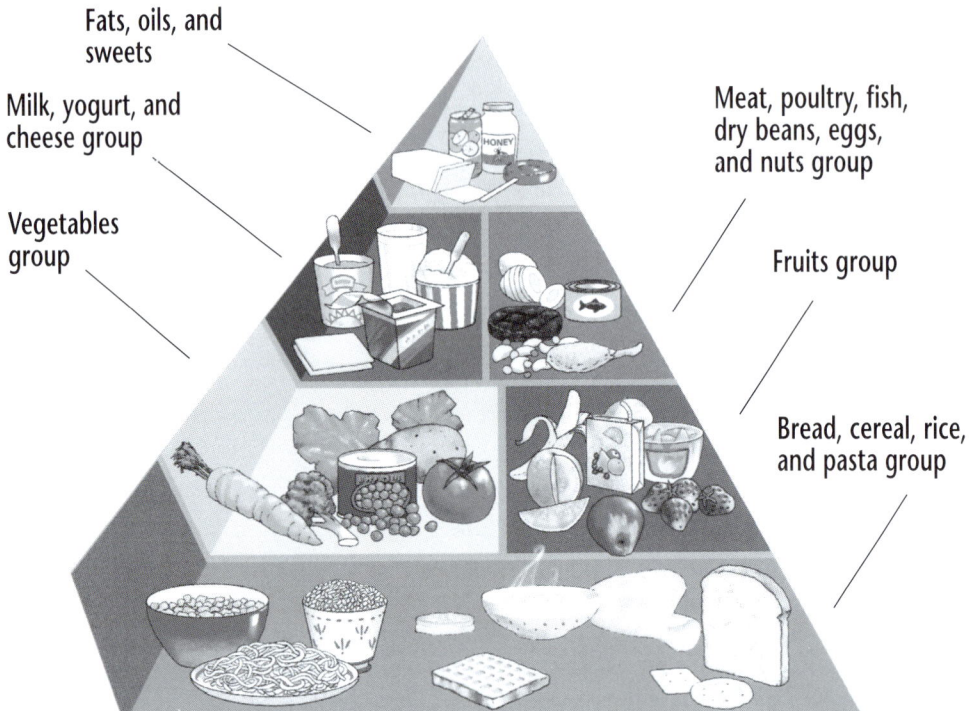

Fats, oils, and sweets

Milk, yogurt, and cheese group

Vegetables group

Meat, poultry, fish, dry beans, eggs, and nuts group

Fruits group

Bread, cereal, rice, and pasta group

The Food Guide Pyramid was created to give you an idea of what to eat each day to maintain a healthful diet.

Guidelines for healthy eating

Try to base your meals on starchy **carbohydrate** foods, such as pasta, rice, and bread, and fruits and vegetables. Most of the **protein** you eat should come from foods such as chicken, fish, and beans.

Eat two to four servings of fruits and three to five servings of vegetables every day. Some examples of one serving include:
- one medium apple, orange, or pear
- 3/4 cup (180 milliliters) of fruit juice
- 3/4 cup (180 milliliters) of vegetable juice
- 1/2 cup (120 milliliters) of fresh or cooked broccoli, carrots, or green beans

Eating a variety of fruits allows you to get the widest range of nutrients.

Fresh fruit and vegetables are best. They have more **vitamins** than canned or frozen fruit. Peaches, melons, berries, green beans, and asparagus are in season in the summer. Apples, pears, oranges, white potatoes, and sweet potatoes are good to eat in the winter.

Balance fats so you eat mainly **polyunsaturated** or **monounsaturated** fats from foods such as oily fish, nuts, seeds, and vegetable oils. Try not to eat foods that are high in **saturated** fats, such as chocolate, cookies, and potato chips, very often.

Even though we should not eat too much fat, fats play an important role in a healthful diet. There are "good fats" as well as "bad fats." It is up to us to see that we get the balance right for our bodies.

Glossary

amino acid smaller unit or building block of proteins. Different amino acids combine together to form a protein.

antioxidant type of vitamin or substance believed to protect body cells from damage and aging

artery tube that carries blood from the heart to different parts of the body

atom smallest part of any element

bacteria microscopic living things. Some are helpful, like those in our intestines, but some can cause disease.

basal metabolic rate (BMR) rate at which the body uses energy when it is at rest

bile substance made in the liver that breaks up the fats in food

blood clot thick mass of blood

bran outer layer of a grain

calorie measurement of energy supplied by food

carbohydrate substance in food that the body uses to provide energy. Foods rich in carbohydrates include bread, rice, potatoes, and sugar.

carbon one of the most common elements, or simple substances

cell smallest unit of a plant or animal

chemical reaction when two or more chemicals react together to produce a change

cholesterol fatty substance found in some foods and in most parts of your body, including the blood

chyme mushy liquid that passes from the stomach to the small intestine

developed country wealthy country that has well-established industries and services

developing country poorer country that does not have well-established industries or services

diabetes mellitus disease in which the body cannot control the level of sugar in the blood

digestive system all the parts of the body that are used to digest food

dissolve break down or mix with a liquid so that the liquid is the same throughout

element chemical that is made up of atoms that are all the same type. Carbon, oxygen, and hydrogen are three elements.

emulsification process of distributing small drops of fat throughout a liquid

energy ability to do work or to make something happen

enzyme substance that helps a chemical reaction take place faster

essential fatty acids (EFAs)
fatty acids that the body needs
but can only get from food

fat soluble can **dissolve** in fats

fatty acid kind of acid found
in animal fat and vegetable oils
and fats

fiber substance found in plants
that cannot be digested by the
human body

free radical chemical in the
body that may harm health

gene information in the form
of a body chemical, DNA, which
carries the instructions for a
living thing to develop and
survive

germ (of wheat or corn)
central part of grain that
contains oil

glucose simple form of sugar
that is broken down from
carbohydrate food during
digestion

glycerol simple substance that
is part of fats

glycogen substance made
from glucose that is stored in the
liver and muscles following
absorption

HDL cholesterol fatty
substance carried in the blood
by "high density lipoproteins"
that help reduce risk of heart
disease

hormone substance made by
different glands in the body that
affects or controls certain organs,
cells, or tissues

hydrogen invisible gas that is
one of the gases in the air.
Hydrogen combines with other
substances to form, for example,
water, sugar, proteins, and fats.

hydrogenation process used
to make fats solid

immune system body's
defense against germs and
diseases

immunized protected against
disease by a vaccination

insoluble cannot be dissolved
in liquid

insulin hormone that controls
the amount of sugar in the
blood

LDL cholesterol fatty
substance carried in the blood
by "low density lipoproteins." A
large amount circulating in the
blood increases the risk of heart
disease.

liver organ in the body that
plays a role in digestion. It
makes bile and helps clean the
blood. People also eat beef and
chicken livers, which are a rich
source of vitamins and minerals.

metabolize to use to gain
energy or build new body tissues

mineral nutrient found in
foods that the body needs to
stay healthy

molecule   smallest unit of a substance that is still that same substance and still has the same properties as the substance

monounsaturated   fatty acid that has one pair of hydrogen atoms missing

nutrient   substance found in foods that helps the body grow and stay healthy. Proteins, carbohydrates, fats, vitamins, and minerals are all nutrients.

obese   describes a person whose weight is twenty percent or more above a healthy weight

omega-3 fatty acids   group of essential fatty acids found in soybeans, fish, walnuts, and dark-green, leafy vegetables

omega-6 fatty acids   group of essential fatty acids found in vegetable fats

organ   body part that has a special job to do

oxygen   gas present in the air and used by the body. Oxygen is one of the most common elements

pancreas   gland that produces various digestive juices that flow through a tube into the small intestine

polyunsaturated   fatty acid that has more than one pair of hydrogen atoms missing

processed   describes foods that have been changed to make them easier to prepare or cook

protein   complex chemical that the body needs to grow and repair cells

puberty   time of life when a child's body develops into an adult's body

saliva   watery liquid made by glands in the mouth and the inside of the cheeks

saturated   fatty acid that is saturated with hydrogen atoms

solution   liquid that has a substance dissolved in it

soluble   able to dissolve in liquid

starch   carbohydrates stored in plants

stroke   sudden change in blood supply to the brain that can cause loss of movement in parts of the body

trans fat   type of fat made by hydrogenation

tissue   material made up of cells that forms a part of an animal or of a plant

triglyceride   chemical form in which fats exist in foods and in the body

unsaturated   fatty acid that is not saturated with hydrogen atoms

virus   tiny, nonliving thing inside your body that makes you sick

vitamin   nutrient needed by the body in small amounts

Further Reading

Akers, Charlene. *Obesity.* San Diego, Calif.: Lucent Books, 2000.

D'Amico, Joan, and Karen Eich Drummond. *The Healthy Body Cookbook.* Hoboken, N.J.: John Wiley & Sons, 1999.

Kalbacken, Joan. *The Food Pyramid.* Danbury, Conn.: Children's Press, 1998.

Kreitzman, Sue. *Low Fat Cookbook.* New York: Doring Kindersley Publishing, 2000.

Petrie, Kristin. *Fit and Fats.* Edina, Minn.: ABDO Publishing, 2003.

Royston, Angela. *Eating and Digestion.* Chicago: Heinemann Library, 1998.

Thomas, Ann. *Fats, Oils, and Sweets.* Bromall, Pa.: Chelsea House Publishers, 2003.

Toriello, James. *The Stomach: Learning How We Digest.* New York: Rosen Publishing, 2001.

Weintrab, Aileen. *Everything You Need to Know about Eating Smart.* New York: Rosen Publishing, 2000.

Westcott, Patsy. *Diet and Nutrition.* Austin, Tex.: Raintree Publishers, 2000.

Index